Careers without College

Hazardous Waste Technician

by David Heath

Consultant:

Andrew J. McCusker
President, Mackworth Environmental Management
President, National Association
of Environmental Professionals

CAPSTONE
HIGH/LOW BOOKS
an imprint of Capstone Press
Mankato, Minnesota

MESA VERDE MIDDLE SCHOOL
8375 Entreken Way
San Diego, CA 92129

Capstone High/Low Books are published by Capstone Press
818 North Willow Street, Mankato, Minnesota 56001
http://www.capstone-press.com

Copyright © 1999 Capstone Press. All rights reserved.
No part of this book may be reproduced without written permission from the publisher.
The publisher takes no responsibility for the use of any of the materials or methods described in this book, nor for the products thereof.
Printed in the United States of America.

Library of Congress Cataloging-in-Publication Data
Heath, David, 1948–
  Hazardous waste technician/by David Heath.
  p. cm.—(Careers without college)
  Includes bibliographical references (p. 44) and index.
  Summary: Outlines the educational requirements, duties, salary, employment outlook, and possible future positions for hazardous waste technicians.
  ISBN 0-7368-0171-5
  1. Hazardous waste management—Vocational guidance—Juvenile literature.
[1. Hazardous waste management—Vocational guidance. 2. Vocational guidance.]
I. Title. II. Series: Careers without college (Mankato, Minn.)
TD1030.5.H43 1999
628.4'2'023—dc21   98-51610
                                                                                       CIP
                                                                                        AC

Editorial Credits
Leah Pockrandt, editor; Steve Christensen, cover designer; Kimberly Danger and
    Sheri Gosewisch, photo researchers

(The affiliation of the consultant with the National Association of Environmental Professionals does not indicate the endorsement of this book by the National Association of Environmental Professionals.)

Photo Credits
Harding Lawson Associates, 30
Index Stock Imagery/Tibor, 34
International Stock/P. K. Trace, 16; Vincent Graziani, 29
Mark Ide, 19, 26
Mary E. Messenger, 20, 22
Photo Network/Stephen Agricola, 4, 12; Mark Sherman, 9, 14, 24, 32, 38, 47
Photophile/Mark E. Gibson, 36
Shaffer Photography/James L. Shaffer, 11
Unicorn Stock/Joel Dexter, 41
Uniphoto, cover; Mellott, 6

Table of Contents

Fast Facts 5

Chapter 1 Job Responsibilities 7

Chapter 2 What the Job Is Like 17

Chapter 3 Training..................... 25

Chapter 4 Salary and Job Outlook 33

Chapter 5 Where the Job Can Lead 37

Words to Know 42

To Learn More 44

Useful Addresses 45

Internet Sites 46

Index 48

Fast Facts

Career Title — Hazardous waste technician

Minimum Educational Requirement — U.S.: high school diploma
Canada: high school diploma

Certification Requirement — U.S.: varies by state
Canada: none

Salary Range — U.S.: $18,000 to $40,000
(Information based on industry estimates, late 1990s figures)
Canada: $18,000 to $60,000
(Canadian dollars)

Job Outlook — U.S.: good
(Information based on industry estimates, late 1990s projections)
Canada: good

DOT Cluster — Professional, technical, and managerial occupations
(Dictionary of Occupational Titles)

DOT Number — 168.267-086

GOE Number — 11.10.03
(Guide for Occupational Exploration)

NOC — 2263
(National Occupational Classification—Canada)

Chapter 1

Job Responsibilities

Hazardous waste technicians move, store, and dispose of hazardous waste. Items such as asbestos, lead, and household cleaners can make people ill. Some even can cause death. Hazardous waste also can harm the environment. Hazardous materials must be disposed of properly.

Hazardous waste includes items that are flammable, corrosive, explosive, or toxic to the environment or people. Flammable items are likely to catch fire. Corrosive items react dangerously with other materials.

Hazardous waste technicians clean up dangerous materials. These technicians pack harmful substances in safe containers. They also transport this hazardous

Hazardous waste technicians must properly handle waste materials.

waste from one place to another. Technicians make sure the waste is properly disposed of or recycled.

Safety First

Hazardous waste technicians are concerned about their personal safety. Technicians learn how to work safely.

Hazardous waste technicians use special safety equipment. For example, they may wear respirators. This breathing equipment allows technicians to work around dust or dangerous fumes. Technicians often wear special clothing and gloves that protect them from chemicals. Most protective suits are made of a synthetic material. This manufactured material is light, waterproof, and resistant to chemicals. Face masks or goggles help protect technicians' eyes and skin.

Hazardous waste technicians must know about a variety of dangerous materials. Each type of hazardous waste must be treated differently. Technicians know the safety procedures for different kinds of waste. Technicians also know how to handle materials so they will not pollute the soil, water, or air.

Hazardous waste technicians must carefully store harmful materials. Technicians must transport these

Hazardous waste technicians must carefully store hazardous materials.

materials to special places for disposal. Technicians store some chemicals in large steel or plastic drums. Technicians store other hazardous materials in cardboard cartons lined with plastic. These storage containers keep waste materials safe during transportation.

Some waste materials are overpacked. Technicians place these materials in two containers in case one becomes damaged. But wastes such as paints may be left in their original containers for disposal.

Hazardous waste technicians work in a variety of settings. Some technicians work on sites to clean up hazardous materials in soil and groundwater. Groundwater is water within the ground that supplies wells and springs. Other technicians mainly help clean up vacant buildings. Still others work with hazardous materials from ongoing operations such as chemical or petroleum processing.

Procedures in Handling Hazardous Waste
Hazardous waste technicians follow special procedures to dispose of hazardous waste. These processes enable technicians to handle the waste materials safely. There are a variety of ways to rid an area of hazardous waste. For example, the U.S. government has approved more than 20 different methods to dispose of hazardous waste.

Project managers usually first establish what types of hazardous substances are at project sites. Project

Hazardous waste technicians must know how to handle a variety of waste materials.

managers are in charge of these sites. Project managers then tell hazardous waste technicians what substances they must clean up at these sites. Technicians follow the directions of project managers, site managers, or other supervisors.

Hazardous waste technicians use a variety of equipment in their work.

Hazardous waste technicians clean up the dangerous substances. They may use shovels, buckets, and scrapers. They may use special vacuum cleaners, power equipment, chemical analyzers, or detection meters at project sites. Chemical analyzers indicate what level of certain chemicals are present.

These levels tell technicians if a particular site needs to be cleaned up. The levels also indicate if the technicians have the right protective clothing and equipment. Detection meters indicate if dangerous gases are in the air.

Hazardous waste technicians next pack the hazardous materials in special containers. Technicians make sure the materials are safe for transport. Technicians then haul hazardous materials to warehouses in trucks.

At warehouses, hazardous waste technicians sort through the materials. They separate different kinds of waste. Some solid materials are shipped to landfills. Asbestos is one such material. Asbestos was once a popular building material. Breathing its fibers is now known to cause serious illness. Technicians seal liquid materials such as insecticides in large barrels. Insecticides are chemicals used to kill insects. Barrels may be stored in warehouses or buried in concrete.

Many hazardous waste products can be recycled. Hazardous waste technicians send these materials to processing plants. Technicians filter

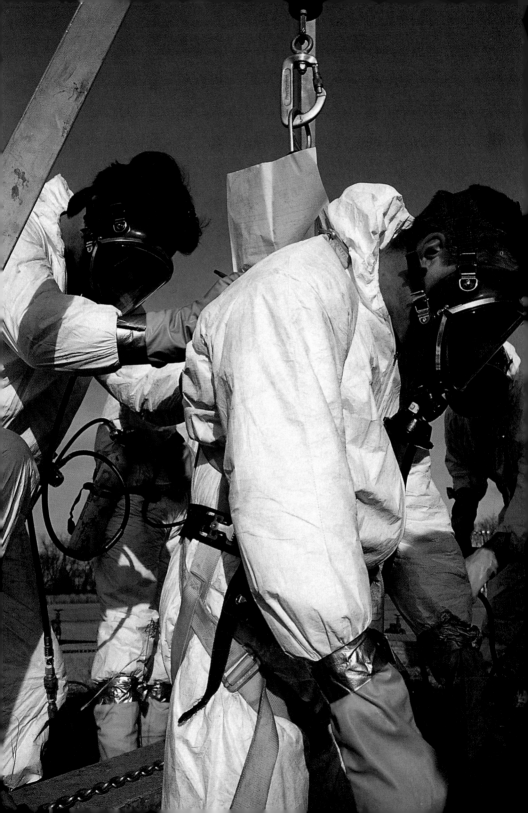

automotive fluids such as oil and antifreeze. These liquids can be used again after they are strained through screens. Latex paint is dried out. The latex then is used to make caulk. This waterproof paste is used to seal seams on windows and doors.

Specialties

Hazardous waste technicians can specialize in handling certain materials. They may become experts at the removal of certain waste products. For example, some technicians specialize in the removal of asbestos. Other technicians specialize in handling radioactive waste. Nuclear power plants and nuclear weapons production create radioactive waste. Medical research and patient treatment also produce radioactive waste.

Hazardous waste technicians have other specialties as well. They may handle substances such as medical waste, household chemicals, and lead. Hospitals and clinics produce a variety of medical waste. This waste includes used needles, bandages, and blood samples. This waste can make people sick. Medical waste needs to be disposed of properly.

Hazardous waste technicians may become experts at the removal of certain waste products.

Chapter 2
What the Job Is Like

Hazardous waste technicians work with many other people. Several technicians may work together on projects. Technicians work under the guidance of project managers. Technicians also work with environmental engineers. These engineers plan how to protect the environment. They sometimes determine how hazardous waste sites will be cleaned up.

Hazardous waste technicians also work with many kinds of scientists. Technicians may gather soil and water samples for chemists or other scientists to test. Scientists test the samples to see if they contain hazardous materials. Technicians also may assist scientists in laboratory tests.

Various kinds of scientists may test samples gathered by hazardous waste technicians.

Work Settings

Hazardous waste technicians work in a variety of settings. For example, they may work at facilities where people drop off household waste. They may work at hazardous waste recycling centers. Technicians may work at factories where waste is produced. They also may work in warehouses where waste is stored. But hazardous waste technicians do most of their work in the outdoors. Most hazardous waste that needs to be collected is outdoors.

Most hazardous waste technicians work during the day. But some technicians work varied hours. These individuals may work day or night shifts. Most technicians work eight-hour shifts. But they may have to work extra hours during emergencies. For example, technicians may work extra hours to clean up a chemical or oil spill. Technicians may be required to do dangerous work at night with few people around.

Governing Organizations

Hazardous waste technicians must follow rules to handle and dispose of hazardous materials. Technicians must know how to apply these rules to

Hazardous waste technicians must be available when emergencies occur.

---

different situations. The United States and Canada each has three government agencies that make rules about hazardous materials. In Canada, hazardous materials are called special waste.

In the United States, the Environmental Protection Agency (EPA) makes rules about air, water, and land quality. This government organization also governs the

Hazardous waste technicians must follow safety rules.

storage and disposal of hazardous waste. The EPA makes sure companies handle hazardous materials properly. The EPA checks factories to make sure they do not dump hazardous waste in unsafe places. The EPA decides what areas may need to be cleaned up. It also monitors the clean-up activities.

In Canada, individual provinces make rules regarding the environment and hazardous waste. The Ministry of the Environment in each province makes and enforces these rules.

In the United States, the Occupational Safety and Health Administration (OSHA) deals with hazardous substances at workplaces. Hazardous waste technicians must follow OSHA rules. These rules help technicians work safely. OSHA rules tell what safety equipment technicians must wear. OSHA rules also control how hazardous materials should be handled.

In Canada, two agencies deal with hazardous substances at workplaces. Labour Canada oversees safety in federal workplaces such as banks and airports. The Ministry of Labour and Workers' Compensation Board in each province oversees safety in all other workplaces.

In the United States, the Department of Transportation (DOT) makes rules about how hazardous materials can be moved from place to place. Technicians need to record what kind of materials are shipped. They also need to record

Hazardous waste technicians should be in good physical shape.

where the materials are shipped. In Canada, Transport Canada mainly governs these actions.

Personal Qualities

Hazardous waste technicians must be hard workers. Many hazardous waste sites must be cleaned up quickly to protect people and the environment. Technicians must work until this clean-up is done.

Hazardous waste technicians should be healthy and in good physical shape. Technicians sometimes dig with shovels or move heavy barrels filled with chemicals. They may haul heavy equipment to waste sites. Technicians must wear protective clothing. They also work in a range of weather and conditions. These factors can cause stress on their bodies.

Hazardous waste technicians must work well in dangerous situations. They must pay attention to all safety procedures. Technicians also must be able to work under pressure. They must remain calm and careful in order to stay safe.

Hazardous waste technicians should be interested in science. Technicians need to know how different chemicals will react with each other. They also need to know how chemicals will affect the environment.

Hazardous waste technicians need good communication skills. They need to listen carefully to project managers and others. They need to understand directions well. They also need to talk to project managers and other technicians about their work.

Chapter 3

Training

Both the United States and Canada have basic training requirements for hazardous waste technicians. Many technical schools, colleges, and universities offer hazardous waste management courses. But hazardous waste technicians learn many of their skills on the job.

What Hazardous Waste Technicians Study

Hazardous waste technicians in the United States need a high school diploma. In Canada, most hazardous waste technicians have a high school diploma. But a high school diploma is not required in Canada. Technicians in both the United States and Canada also may take some hazardous waste management courses.

Hazardous waste technicians learn many of their skills on the job.

Hazardous waste technicians must complete basic hazardous materials training.

Licensing and Certification

Hazardous waste technicians must have some basic training in both the United States and Canada. In the United States, technicians must complete a basic training program to fulfill OSHA requirements.

Most states require hazardous waste technicians to be certified in certain areas of work. Technicians take classes at technical schools, community colleges, or universities to earn certificates. The certificates show that technicians are trained to handle certain kinds of hazardous waste. But technicians in Canada are not required to get certificates after their training.

Hazardous waste technicians in the United States also may earn certificates for transporting hazardous materials. These certificates show that technicians know the proper ways to store and move materials.

In the United States, companies that hire hazardous waste technicians must be licensed by the government. Licenses show that these companies follow proper procedures for handling hazardous materials. The companies also need to train their technicians in these procedures. Companies must follow all OSHA job-safety rules to keep their licenses. They also must follow the rules of the EPA and DOT.

In Canada, companies may be required to obtain permits or licenses. Companies that produce hazardous waste must register with the province's Ministry of the Environment. Organizations that recycle, treat, store, or transport special waste must obtain permits or licenses. These items also come from the Ministry of the Environment for the province. Environment Canada may require facilities to obtain permits where federal regulations exist. Environment Canada is a science-based agency that regulates industries that affect the environment.

What You Can Do Now

Persons interested in becoming hazardous waste technicians should take science classes in high school. Chemistry is one of the most important science classes for students interested in this career. Technicians are better prepared if they understand how chemicals react with one another.

Earth science and environmental science also can be helpful. These classes will help students understand how chemicals affect the land, water,

Companies that produce hazardous waste must register with state or provincial agencies.

and air. Biology teaches students how chemicals or particles in the air or water can affect basic processes. It also helps students understand how exposure to chemicals can affect humans and animals.

Hazardous waste technicians must be able to understand and follow instructions.

Students who want to be hazardous waste technicians can benefit from other subjects as well. For example, communication classes help prepare students for this career. Technicians need to read and write well. They must understand instructions and talk with others. Some technicians must write reports. They also often are required to read detailed technical information.

Other activities also can teach students the effects of hazardous waste on the environment. For example, students can read books and magazines about the environment and how chemicals can affect human health. They also can join environmental clubs.

People interested in becoming hazardous waste technicians may benefit from learning how to handle machinery. Some technicians drive trucks and operate heavy equipment. For example, technicians may drive forklifts in storage warehouses. Technicians may use these vehicles to pick up and move heavy items. Some technicians use a special type of forklift called a clamp jeep. This piece of equipment only is used to pick up drums.

Chapter 4

Salary and Job Outlook

Several factors affect the salaries that hazardous waste technicians receive. These factors include work location, education, and experience.

Hazardous waste technicians in the United States earn between $18,000 and $40,000 per year (all figures late 1990s). But the average salary is $25,000 to $30,000 per year.

In Canada, hazardous waste technicians' salaries vary according to where they work. Technicians who work for government agencies earn between $18,000 and $30,000 per year. Technicians who work for private companies earn

Work location can affect the salaries of hazardous waste technicians.

Trained hazardous waste technicians will continue to be needed.

between $35,000 and $60,000 per year. Most technicians in Canada work for private companies.

Most hazardous waste technicians are paid by the hour. The number of hours they work determines their salaries. Some technicians may work more than 40 hours per week. The extra hours are called overtime. These technicians are paid for their extra work. This makes technicians' salaries vary.

Benefits
Hazardous waste technicians usually receive benefits from their employers. Benefits can include insurance, retirement funds, and paid sick and vacation time. Money in retirement funds helps support technicians when they retire.

Job Outlook
The hazardous waste field has grown greatly. Companies continue to produce hazardous waste. Other hazardous waste already exists in many places. Trained technicians are needed to properly dispose of and store these hazardous materials.

    Government agencies also hire hazardous waste technicians. These agencies hire technicians to clean up areas where hazardous waste was disposed of improperly.

Chapter 5

Where the Job Can Lead

Hazardous waste technicians can advance in many ways. They can advance within companies or agencies through experience. They can earn advanced degrees in their fields. They also can move to other companies for better jobs. Some technicians can advance to jobs in related careers.

Advancement

Hazardous waste technicians can advance by gaining experience. For example, they can learn to use special equipment such as chemical analyzers.

Hazardous waste technicians can advance by gaining experience.

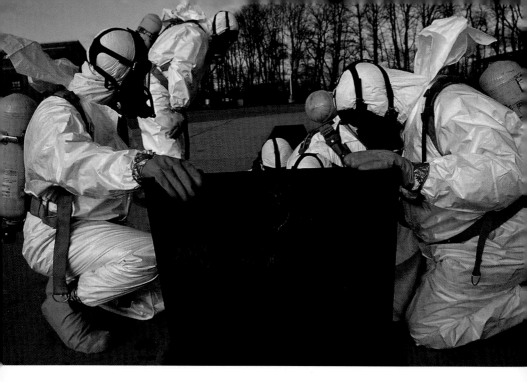

Hazardous waste technicians can advance by continuing their educations.

Hazardous waste technicians who gain experience may become team leaders. These individuals lead groups of technicians. Team leaders assign work and oversee the work of other hazardous waste technicians.

Hazardous waste technicians also can advance by continuing their educations. They can earn certificates for handling special hazardous materials. They can earn degrees in waste management. Technicians who have some college training often earn better salaries.

Hazardous waste technicians with advanced degrees may become project managers. These individuals usually have a bachelor's degree. People earn a bachelor's degree by completing a course of study at a college or university. People usually earn a bachelor's degree in about four years. Project managers typically have degrees in chemistry, environmental science, or a type of engineering. Project managers also may have a degree in earth science or geology.

Some people who work in the hazardous waste industry have a master's degree. People earn a master's degree by completing an advanced course of study at a college or university. People can earn a master's degree in about two years. Additional education often increases people's chances to advance in their careers.

Related Careers

Other careers are related to hazardous waste management. For example, people interested in hazardous waste management may become environmental inspectors. These individuals make sure EPA rules are obeyed. Inspectors often work for city, county, state, or federal government agencies.

People who work in the sanitation industry have tasks similar to hazardous waste technicians. For example, sanitation workers transport waste. Workers at water treatment plants make sure water is processed properly.

People interested in waste management and science can become scientists, engineers, or chemists. These individuals may find ways to better handle hazardous materials.

People interested in waste management can become chemists, engineers, or scientists.

The opportunities in hazardous waste management will continue to increase. Companies will produce more hazardous materials. Sites will need to be cleaned. Trained individuals will be needed to properly store, transport, and dispose of hazardous waste.

Words to Know

asbestos (ass-BESS-tuhss)—a grayish mineral whose fibers are woven into a fireproof fabric; asbestos once was a popular building material; asbestos rarely is used now because people who breathe its fibers can become seriously ill.

chemical (KEM-uh-kuhl)—a substance used in or produced by chemistry; medicines, gunpowder, and household cleaners all are made from chemicals.

chemistry (KEM-is-tree)—the scientific study of substances including what they are made of and how they react with each other

hazardous waste (HAZ-ur-duhss WAYST)—dangerous materials that need to be disposed of safely

hazardous waste site (HAZ-ur-duhss WAYST SITE)—a place where dangerous materials have been disposed of

inspector (in-SPEK-tur)—a person who checks or examines facilities and hazardous waste sites

radioactive waste (ray-dee-oh-AK-tiv WAYST)—materials produced by the creation of nuclear weapons and power plants; radioactive materials give off harmful radiation that can cause illness or death.

respirator (RESS-puh-ray-tur)—a device worn over the mouth or nose that cleans the air so a person can breathe safely

To Learn More

Fanning, Odom. *Opportunities in Environmental Careers.* VGM Opportunities Series. Lincolnwood, Ill.: VGM Career Horizons, 1996.

Field, Shelly. *100 Best Careers for the 21st Century.* New York: MacMillan, 1996.

Gano, Lila. *Hazardous Waste.* Lucent Overview Series. Our Endangered Planet. San Diego: Lucent Books, 1991.

Gold, Susan Dudley. *Toxic Waste.* Earth Alert. New York: Crestwood House, 1990.

Kowalski, Kathiann M. *Hazardous Waste Sites.* Minneapolis: Lerner, 1996.

Useful Addresses

Environment Canada Enquiry Center
351 St. Joseph Boulevard
Place Vincent Massey
Hull, QC K1A 0H3
Canada

Institute for Hazardous Materials Management
11900 Parklawn Drive
Suite 450
Rockville, MD 20852

National Association of Environmental Professionals
6524 Ramoth Drive
Jacksonville, FL 32226-3202

U.S. Environmental Protection Agency
401 M Street SW
Washington, DC 20460-0003

Internet Sites

Environment Canada's Green Lane
http://www.ec.gc.ca

HAZMAT Safety
http://hazmat.dot.gov

Illinois Waste Management and Research Center
http://www.wmrc.uiuc.edu

Institute of Hazardous Materials Management
http://www.ihmm.org

National Association of Environmental Professionals
http://www.naep.org

U.S. Environmental Protection Agency
http://www.epa.gov

Index

advancement, 37, 39–40
asbestos, 7, 13, 15

benefits, 35

certification, 26
chemical analyzers, 12, 37
chemists, 17, 40

Department of Transportation (DOT), 21, 27
detection meters, 12

Environmental Protection Agency (EPA), 19–20, 27, 40

Labour Canada, 21

Ministry of Labour and Workers' Compensation Board, 21

Ministry of the Environment, 21, 28

Occupational Safety and Health Administration (OSHA), 21, 26, 27
outlook, 35

project managers, 10, 11, 17, 23, 39

radioactive waste, 15
respirators, 8

salary, 33–35
scientists, 17, 40
specialties, 15

training, 25–27, 39
Transport Canada, 21